MIX
Papier aus verantwortungsvollen Quellen
Paper from responsible sources
FSC® C105338

Md Belal Bin Heyat

Insomnia: Medical Sleep Disorder & Diagnosis

Anchor Academic Publishing

Heyat, Md Belal Bin: Insomnia: Medical Sleep Disorder & Diagnosis, Hamburg, Anchor Academic Publishing 2017

Buch-ISBN: 978-3-96067-089-6
PDF-eBook-ISBN: 978-3-96067-589-1
Druck/Herstellung: Anchor Academic Publishing, Hamburg, 2017

Bibliografische Information der Deutschen Nationalbibliothek:
Die Deutsche Nationalbibliothek verzeichnet diese Publikation in der Deutschen Nationalbibliografie; detaillierte bibliografische Daten sind im Internet über http://dnb.d-nb.de abrufbar.

Bibliographical Information of the German National Library:
The German National Library lists this publication in the German National Bibliography. Detailed bibliographic data can be found at: http://dnb.d-nb.de

All rights reserved. This publication may not be reproduced, stored in a retrieval system or transmitted, in any form or by any means, electronic, mechanical, photocopying, recording or otherwise, without the prior permission of the publishers.

Das Werk einschließlich aller seiner Teile ist urheberrechtlich geschützt. Jede Verwertung außerhalb der Grenzen des Urheberrechtsgesetzes ist ohne Zustimmung des Verlages unzulässig und strafbar. Dies gilt insbesondere für Vervielfältigungen, Übersetzungen, Mikroverfilmungen und die Einspeicherung und Bearbeitung in elektronischen Systemen.

Die Wiedergabe von Gebrauchsnamen, Handelsnamen, Warenbezeichnungen usw. in diesem Werk berechtigt auch ohne besondere Kennzeichnung nicht zu der Annahme, dass solche Namen im Sinne der Warenzeichen- und Markenschutz-Gesetzgebung als frei zu betrachten wären und daher von jedermann benutzt werden dürften.

Die Informationen in diesem Werk wurden mit Sorgfalt erarbeitet. Dennoch können Fehler nicht vollständig ausgeschlossen werden und die Diplomica Verlag GmbH, die Autoren oder Übersetzer übernehmen keine juristische Verantwortung oder irgendeine Haftung für evtl. verbliebene fehlerhafte Angaben und deren Folgen.

Alle Rechte vorbehalten

© Anchor Academic Publishing, Imprint der Diplomica Verlag GmbH
Hermannstal 119k, 22119 Hamburg
http://www.diplomica-verlag.de, Hamburg 2017
Printed in Germany

ACKNOWLEDGEMENT

The delight of delving deep into the intricacies of engineering research is immense and it surmounts the ecstasy of reaching the destination. It was ended rapturous to dive through the waters of engineering research and this fascinating task becomes all the more blissful due to the presence of two scintillating stars with me: my supervisor and coordinator.

During the entire course of this study, my supervisor **Er Mohd Maroof Siddiqui** and **Prof. (Dr) Syed Hasan Saeed** has been a great source of encouragement and inspiration, knowledge and wisdom without whose patronage this work could not have seen the light of the day. I feel privileged to extend my profound thanks to my cousin, **Er Arshad Mehdi** who really showed me the path, blessed me to boost of my morale, and motivated me to see the work in its present form. I feel really incapable of expressing my heartfelt and grateful thanks to my reverent teachers: **Er Serajul Haque,** Deptt of Mechanical Engineering, King Khalid University, KSA, **Prof. (Dr) Hasin Alam,** Deptt of ECE, Ibra College of technology, Oman, **Prof. (Dr) H.H.Siddiqui,** Dean Deptt of Pharmacey and Advisor to VC, **Er Naseem Ahmad,** Deptt of Electrical Engineering, **Er Manoj Kumar Singh, Prof. Aqueel Ahmad,** Deptt of Cvil Engineering, **Dr Khawaza Moeed,** Principal, University Polytechnic, **Dr VS Chandel, Dr Asim Siddiqui,** Deptt of Physics, **Dr Abdul Rahman Khan,** HOD, **Dr Zulfiqar Ali,** Deptt of Chemistry, **Dr Azizur Rahman,** Deptt of Mathematics, **Er Mani Rajput, Prof. Naimur Rahman Kidwai** and **Er Saima Beg,** Deptt of ECE, Integral University, Lucknow, for their encouragement. I could not forget to offer to thanks to my teacher **Munawar Alam** and **Shailendra Singh** Deptt of ECE, Integral University, Lucknow, who with their talent and experience, guided me out all hurdles.

"Keep thy friend, Under thy own life's key" – **William Shakespeare.**

I express heartiest thanks to all my friends, especially **Faijan Akhtar, Shahnawaz, Gaurav Singh, Seema Rajmani,** *Shaguftah***, Sophiya Azim Khan, Shammi Waseem, Tanya Gupta** and **Mansoor Hasan Yasir.** Right from the inception of my present study to finally the crystallization of its results my parents, **Dr Mohd Sikandar Hayat Siddiqui** and **Mrs Roshan Ara** have stood by me through thick and thin my father's association with sleep disorder and his passion descended on me from everywhere at home, elating of my spirits and clouring of my research. My mother had taught me to construct sentences as a school kid. She must be a happy person today, seeing me grow big enough to write a thesis. Their constant encouragement endowed me with the necessary push up to demystify the obstacles coming in the way. Equally matchless was the contribution of my brother **Mohd Ammar Bin Hayat** and my cousin **Sumbul Mehdi,** who has helped me successed in converting my long cherished aspiration into an everlasting entity. Without his help I would not have been able to operate the complicated statistical software and tabulate the data. My vocabulary falls short when it comes to my cousin **Shafan Azad**. His cooperation in computer work his worthy of applause he has a knack for cheering me during tough times. I cannot forget to thanks my cousin **Shadab Azad** and my aunty **Shahnaj Ara** to help & support of my thesis. My cousin **Shajaan Azad**'s valuable suggestion gave me an opportunity to improve the present work.

Md Belal Bin Heyat

ABSTRACT

Sleep is an important phenomenon in everyone life. A person spends one-third of his life in sleep. Lack of sleep may result into several sleeping disorders. These sleep disorders can affect mental, emotional as well as physical well-being of a person. In this research work, a similar disease known as Insomnia has been discussed. In a normal person, dreaming is an activity which is running out in his mind and at the same time his body is at rest. Insomnia, or sleeplessness, is a sleep disorder in which there is an inability to fall asleep or to stay asleep as long as desired. A person suffering from an insomnia wake up frequently during the night or wake up early and feels exhausted, slow and not refreshed, with a helplessness to concentrate. The various stages of sleep and the systems affected in the human body have been discussed in detail. The various Insomnia symptoms, its causes, and the treatment are also part of a discussion in this thesis. In this research work, the time-frequency analysis of EEG Signals has been done. The electrical and chemical activities of the brain changes in the presence of any sleep disorder. These changes further affect the waveform of EEG signal. These changes further help in detecting sleep disorders. The use of Short Time-frequency analysis applied on Electroencephalogram (EEG) Signals is made in this research work for diagnosing Insomnia. A comparison between the normalized powers of the two types of patient i.e. normal patient and the Insomnia patient is done to achieve meaningful results.

TABLE OF CONTENTS

ACKNOWLEDGEMENT i
ABSTRACT iii
LIST OF TABLES iv
TABLE OF FIGURES vii
LIST OF ABBREVIATIONS viii
CHAPTER 1 – INTRODUCTION 1
    1.1 Classification of Insomnia 1
        1.1.1 According to etiology 1
        1.1.2 According to sleep pattern 1
        1.1.3 According to duration 2
    1.2 Electroencephalogram 2
        1.2.1 EEG Generation and Brain Physiology 2
        1.2.2 EEG Rhythms 4
        1.2.3 Noise and Artifacts 4
    1.3 Macrostructure of sleep: sleep stages 5
        1.3.1 The waking stage (Stage W) 5
        1.3.2 Stage 1 6
        1.3.3 Stage 2 6
        1.3.4 Stage 3 6
        1.3.5 Stage 4 6
        1.3.6 REM sleep 7
    1.4 Microstructure of sleep 7
        1.4.1 Cyclic alternating pattern (CAP) 7
    1.5 History 8
    1.6 Motivation 10
CHAPTER 2 – LITERATURE REVIEW 11
    2.1 Introduction 11
    2.2 Degree of pineal calcification (DOC) is associated with polysomnographic sleep measures in primary insomnia patients 12
    2.3 Studying Herb-Herb Interaction for Insomnia through the theory of Complementarities 12
    2.4 Internet-delivered or mailed self-help treatment for insomnia? A randomized waiting-list controlled trial 13

2.5 Insomnia Characterization: From Hypnogram to Graph Spectral Theory 13
2.6 Relation Between Heart Beat Fluctuations and Cyclic Alternating Pattern During Sleep in Insomnia Patients 13
2.7 Processing of Signals Recorded Through Smart Devices: Sleep-Quality Assessment.. 14
2.8 Advantages and Challenges of Power Spectral Density Estimation Methods for Scanning Radar Angular Super resolution 14
2.9 Modelling and Control for Simultaneous Laser Beam Alignment of a Dual-PSD Industrial Robot Calibration System 15
2.10 Deep Learning EEG Response Representation for Brain Computer Interface 15
2.11 A New Algorithm for Noise PSD Matrix Estimation in Multi-Microphone Speech Enhancement Based on Recursive Smoothing 16
2.12 Power Spectral Density and High Order Bi-spectral Analysis of Alzheimer's EEG... 16
2.13 Binaural Noise Suppression based on an unbiased estimator of target PSD in Complex Noise Environments 17
2.14 Detection of rapid eye movement behaviour disorder using short time frequency analysis of PSD approach applied on EEG signal (ROC-LOC) 17
CHAPTER 3 – RESEARCH METHODOLOGY 18
3.1 Characteristics of the EEG signal 18
3.2 Experimental Setup 18
    3.2.1 Subject Details 18
3.3 Algorithm 21
CHAPTER 4 – RESULTS AND DISCUSSIONS 25
4.1 Loading the signal 25
4.2 Extracting EEG Signals 25
4.3 Filtering of EEG Signals 26
4.4 Applied Window 26
4.5 Comparison of the Original Wave & Filtered Wave 27
4.6 Estimation of power spectral density 27
4.7 Calculation of normalized power 28
CHAPTER 5 – CONCLUSION 29
FUTURE SCOPE 34
REFERENCES 35
AUTHOR'S PUBLICATIONS 41

LIST OF TABLES

Table 1.1: The summary of the basic human brain waves. 4
Table 3.1: Insomnia Patients Details of duration of start time and end time of 1 minute 19
Table 3.2: Normal Patients Details of duration of start time and end time of 1 minute 20
Table 5.1: Normalized Power of the theta Wave of Normal Patient and Insomnia Patient for FP4-F2 Channel and Stage S1 29
Table 5.2: Normalized Power of the alpha Wave of Normal Patient and Insomnia Patient for FP4-F2 Channel and Stage S1 29
Table 5.3: Normalized Power of the theta Wave of Normal Patient and Insomnia Patient for EMG1-EMG2 Channel and Stage S0 30
Table 5.4: Normalized Power of the alpha Wave of Normal Patient and Insomnia Patient for EMG1-EMG2 Channel and Stage S0 30
Table 5.5: Normalized Power of the theta Wave of Normal Patient and Insomnia Patient for EMG1-EMG2 Channel and Stage S0 31
Table 5.6: Normalized Power of the theta Wave of Normal Patient and Insomnia Patient for EMG1-EMG2 Channel and Stage S1 31
Table 5.7: Normalized Power of the alpha Wave of Normal Patient and Insomnia Patient for EMG1-EMG2 Channel and Stage S1 31
Table 5.8: Normalized Power of the beta Wave of Normal Patient and Insomnia Patient for EMG1-EMG2 Channel and Stage S1 32
Table 5.9: Normalized Power of the theta Wave of Normal Patient and Insomnia Patient for EMG1-EMG2 Channel and Stage REM 32
Table 5.10: Normalized Power of the alpha Wave of Normal Patient and Insomnia Patient for EMG1-EMG2 Channel and Stage REM 33
Table 5.11: Normalized Power of the beta Wave of Normal Patient and Insomnia Patient for EMG1-EMG2 Channel and Stage REM 33

TABLE OF FIGURES

Figure 1.1: Structure of a Neuron 3
Figure 1.2: Diagrammatic representation of Human Brain 3
Figure 1.3: The sleep cycle of a normal healthy adult: the sequence of states and stages of sleep 6
Figure 1.4: Cyclic Alternating Pattern 7
Figure 3.1: Complete EEG Signal 21
Figure 3.2: EEG clip data having DC offset (a) and having DC offset removed (b) 22
Figure 3.3: Welch Power Spectral Density Estimate 24
Figure 4.1: Complete EEG Signal of S0 Stage 25
Figure 4.2: An extracted signal with EMG1-EMG2 25
Figure 4.3: Filtered signals for S0 sleep stage 26
Figure 4.4: Window signals for S_0 Sleep Stage 26
Figure 4.5: Differences between original and filtered waveform 27
Figure 4.6: Estimated PSD of different channels 28

LIST OF ABBREVIATIONS

Ins	Insomnia
N	Normal
RBD	Rapid Eye Movement Behavior Disorder
EEG	Electroencephalography
PSD	Power Spectral Density
t-f	Time- Frequency Analysis
TFD	Time- Frequency Distribution
STFT	Short Time Frequency Analysis
FIR	Finite Response Filter
NA	Stage Not Available
N/A	Channel Not Available
**	Data Not Available
*	Date Not Available
LPF	Low Pass Filter
HPF	High Pass Filter
BPF	Band Pass Filter
BSF	Band Stop Filter
ECG	Electrocardiogram
EMG	Electromyogram
EOG	Electrooculagram
ERG	Electroretinogram
GSR	Galvanic Skin Response
MA	Moving Average
ARMA	Autoregressive Moving Average
AR	Autoregressive
REM	Rapid Eye Movement
NREM	Non-Rapid Eye Movement
DC	Direct Current
CAP	Cyclic Alternating Pattern
EU	Engineering Units
CPAP	Continuous Positive Airway Pressure
ICSD	International Classification Of Sleep Disorders
FFT	Fast Fourier Transform
IFFT	Inverse Fast Fourier Transform

CHAPTER 1
INTRODUCTION

The human mind is single of the supreme compound systems in the world. Nowadays numerous skills exist to high mind waves and EEG is single of them. Insomnia, or sleeplessness, is a sleep disorder in which there is an inability to fall asleep or to stay asleep as long as desired A person suffering from an insomnia wake up frequently during the night or wake up early and feels exhausted, slow and not refreshed, with a helplessness to concentrate. Insomnia is common problem in general population currently. It is similarly single of the main causes of daylight sleepiness. It is also, composed with discomfort and tiredness, the most communal disorder among all of us. However, insomnia is a widespread situation in our humanity; both surgeons and patients are missing in the knowledge about it. There are also no generally accepted standards of treatment.

The effects of insomnia can also include irritable mood and opinion and an increased possibility of accidents while driving or working with machines. Insomnia is not a symptom of other disorders, but it is secondary to other medical conditions

1.1 Classification of Insomnia

1.1.1 According to etiology

Primary insomnia

When insomnia has no identified corporeal (pain), emotional (depression/ anxiety), ecological cause, the situation is baptized primary insomnia.

Secondary or comorbid insomnia

This is once the victim has sleep difficulties because of approximately else, such as a fitness situation like depression, cancer, pain, drug being taken; or a substance being used, like alcohol.

1.1.2 According to sleep pattern

Sleep-onset insomnia

When the sufferer takes a long time to get to sleep, but can sleep concluded the darkness once sleep twitches.

Sleep-maintenance insomnia

When sufferer wakes frequently during the night and sleep is fragmented

1.1.3 According to duration

Transient insomnia

Lasting less than a week, this is the most common and widespread form among the population.

Acute insomnia

Lasting between one and four weeks, it is related to stress issues, but additional longer permanent than for short-lived insomnia.

Chronic insomnia

Lasts for four or more weeks and may be due to intrinsically causes in the organism, eg a long-term physical or psychiatric illness or it may have no apparent underlying cause.

1.2 Electroencephalogram

Electroencephalogram (EEG) reflects the electrical activity occurring at the surface of a functioning brain. Traditionally, EEG is the representation of brain voltage potentials (usually in microvolt) written onto paper. A modern EEG signal is the digitized version of these potentials for computer storage and analysis. At the end of the nineteenth century (1875) Richard Caton, an English physician discovered that the brain of animals produces electricity. In 1924, Hans Berger, a German physiologist registered the first human EEG. Using invasive technique and recorded mostly from his children, Berger's data revealed that human brain produces quasi-sinusoidal oscillation in the wake and relaxed state with the eyes closed (Tong et al. 2009). Unfortunately, it took more than 10 years for the scientific community to accept that the potentials recorded from the human scalp reflect the genuine brain signals. For decades, the clinicians in understanding and treatment of neuro-physiological disorders primarily used EEG. Presently, beside clinical purposes, EEG application has extended to neuroscience, cognition, and in other research fields. Since last decade, EEG is being used for commercial purposes including video game controllers, mind controlled cars, and in numerous other fields, implementing so called brain computer interface.

1.2.1 EEG Generation and Brain Physiology

The neuron is the basic functional unit of the nervous system. The adult human brain consists of approximately 100 billion nerve cells or neurons. A typical neuron consists of a nerve cell body (or soma), short branching fibers (dendrites), and an elongated projection (axon).

Neurons are interconnected into neural nets through synapses. Neurons accept nerve motions (action potentials) and transmit these signals to other neurons.

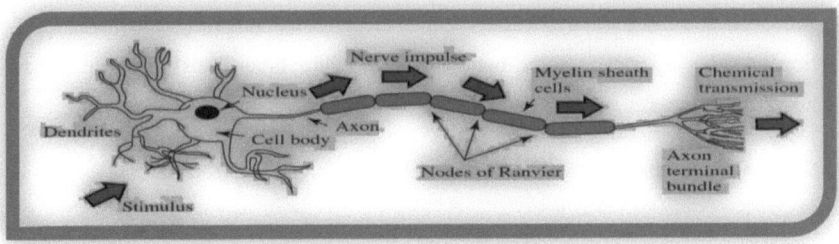

Figure 1.1: Structure of a Neuron

EEG is a blurred and highly attenuated electric potential that results from the activities of multiple groups of neurons from multiple cerebral regions. When neurons are excited, local currents are generated inside the dendrites. The resultant current produces a magnetic field and an electrical field over the brain surface which is measurable by EEG system. The membrane potential governs the direction of the current which is generated by pumping positive ions (Na+, K+ and Ca++) and negative ions (Cl–) through the neuron membrane.

The human brain, from anatomical point of view, can be divided into three parts: the cerebrum (forebrain), the cerebellum (hindbrain), and the brainstem (Figure 1.2). Cerebrum is responsible for regulating higher cognitive and emotional functions (Gerrig et al. 2002). The cerebellum controls voluntary movements of muscles and maintains balance.

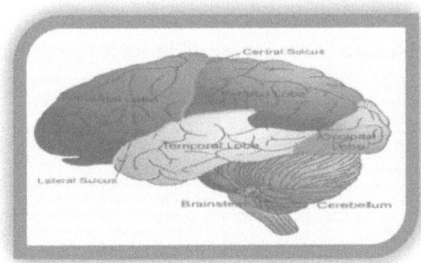

Figure 1.2: Diagrammatic representation of Human Brain

Cerebrum is subdivided into four major lobes: Frontal, Parietal, Temporal, and Occipital. The Frontal lobe is located at the front and associated with working memory, language, social and sexual behavior (Stuss et al. 2002). The Parietal lobe includes processing of tactile sensory information and integration of sensory inputs to the visual stimuli (Macdonald 1953). The Temporal lobe is located on the bottom section of the brain. This lobe is important for emotional responses, interpreting sounds and languages (Larry et al. 2007).

1.2.2 EEG Rhythms

Specific harmonic oscillations commonly called rhythms may be observed in human EEG. Rhythms are classified into five main groups reliant on their incidence ranges. These are alpha (α), beta (β), gamma (γ), theta (θ), and delta (δ). The delta rhythms are of lowest frequency band (range: 0.1-4 Hz) and associated with deep sleep. These rhythms are also found in waking stage. Theta waves (4-8 Hz) appear as consciousness slips towards drowsiness. The alpha waves (8-13 Hz) (introduced by Berger in 1929) are the most important rhythms in the whole gamut of brain waves. Alpha rhythms are found over the occipital region of the brain and indicate relaxed awareness without any concentration. Beta rhythms (14-30 Hz) are the normal working rhythm and associated with active attention, thinking, and solving problems.

Beta waves may be observed in normal adults. The waves with frequency above 30 Hz are known as gamma waves. Amplitudes of gamma waves are low. Although their occurrences are rare, gamma waves are associated with event related synchronization (ERS) and may be used to demonstrate the locus of left and right index finger movement (Pfurtscheller et al. 1994).

Table 1.1: The summary of the basic human brain waves

Wave	Frequency (Hz)	Amplitude (μV)
Delta	0.1-4	100-200
Theta	4-8	Greater than 30
Alpha	8-13	30-50
Beta	13-30	2-20
Gamma	40-80	3-5

1.2.3 Noise and Artifacts

EEG data are subject to various artifacts contaminations. Most of the artifacts stem from the surroundings, subject's body functions, and the experimental equipment or even from within the brain. EEG data, therefore, should be made artifact-free for ease of analysis. The most common artifacts that are found in EEG data are discussed in the following section.

Biological artifacts are the artifacts arising from biological activities. They include eye-blink, muscle, swallowing, facial, and cardio artifacts. Eye-blink introduces large variation both in amplitude and frequency of EEG data, which generally falls in the alpha rhythm. Since alpha is the most important rhythm in EEG, eye-blink artifacts reduce the data classification accuracy substantially (Niedermeyer et al. 1999). Muscle artifacts cause slow drift in amplitude. Facial and eyeball artifacts may also cause distortion in EEG signal. Cardio artifacts arise from the functioning heart of the subject and may be removed using the Electrocardiography (ECG) signal. Unwanted VEP is another biological artifact that can obscure the desired EP in certain experiments.

Beside biological artifacts, other important artifacts, such as AC power line noise, DC noise, etc., may contaminate the EEG signal. AC power line noise (either 50 or 60 Hz) arises from the AC power line in the equipments or surroundings. DC offset may also be frequently observed in the recorded EEG signal. This noise results from the experimental hardware.

It is very often hard to interpret the brain activities by simple visual inspection of scalp EEG, even with substantial training. Application of advanced signal processing, however, may assist in the analysis and separation of correct wave-patterns and the beginnings of EEG mechanisms.

1.3 Macrostructure of sleep: sleep stages

American Sleep Disorders Association considered wakefulness, REM sleep and NREM sleep as main stages of vigilance. After an extensive research based on polysomnography description of sleep, the NREM sleep was divided into four specific stages. These stages are representation of sleep depth. During the night, the frequency of sleep stages alters. SWS dominate in the early hours of sleep while REM sleep usually occurs mainly in the second half of sleep. The five stages of sleep including their repetition, occur cyclically. Healthy human sleep begins in stage 1 sleep and progresses through stages 2, 3, and 4 NREM sleep.

1.3.1 The waking stage (Stage W)

The waking stage is referred to as relaxed wakefulness as it means the body prepares for sleep. Muscles start to relax, and eye movement slows. After closing eyes and relaxing, the EEG typically shows a regular pattern of 8 – 12 Hz, known as alpha waves.

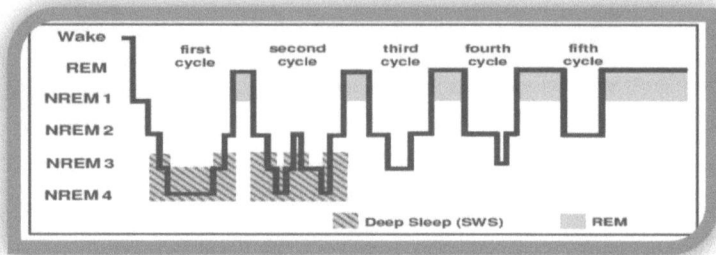

Figure 1.3: The sleep cycle of a normal healthy adult: the sequence of states and stages of sleep

1.3.2 Stage 1

Individual drifts into "Stage 1" sleep and the EEG becomes slower (2 – 7 Hz) and less regular. EEG is reduced in amplitude with little or no alpha (less than 50%). Slow eye movement can be present. EMG level is lower than in the previous stage of waking [14].

1.3.3 Stage 2

Stage 2 is characterized by the appearance of spindles and K-complexes. Spindles are short runs of rhythmical EEG waves of 12 – 16 Hz. Typical K-complexes are EEG waveforms lasting about 0.5 second with a well-delineated negative sharp wave of 12 – 14 Hz, which is immediately followed by a positive component. Should the duration between two following incidences of sleep balusters or K campuses be lower than 3 minutes, this period could be scored as Stage 2, even if there are movement artifacts or increased tonic activity.

1.3.4 Stage 3

Stages 3 is characterized by slow EEG waves of 1 – 2 Hz known as delta waves, which appears in 20%-50% of the period. Also in this stage, sleep spindles and K-complexes may occur.

1.3.5 Stage 4

Same as the Stage 3, the Stage 4 is characterized by slow waves up to 2 Hz which appears in more than 50% of the period.

1.3.6 REM sleep

REM sleep is characterized by an EEG pattern similar to Stage 1, but the rapid eye movements appear on the EEG record and EMG recordings are of lowest amplitude. The heart rate and respiration speed up and become erratic, while the face, fingers, and legs can shudder. Intense visualizing occurs through REM sleep because of heightened cerebral activity.

1.4 Microstructure of sleep

The alternation of the above defined sleep stages represents the macro dynamics of brain. This fluctuation represents micro dynamics of brain. The Cyclic Alternating Pattern (CAP) describes the theory of microstructure of sleep.

1.4.1 Cyclic alternating pattern (CAP)

The cyclic alternating pattern (CAP) is functionally connected with fluctuation of arousal and offers a global framework for characterizing and measuring arousal instability. CAP represents the periodic EEG activity of NREM sleep, characterized by sequences of different transient electrocortical events.

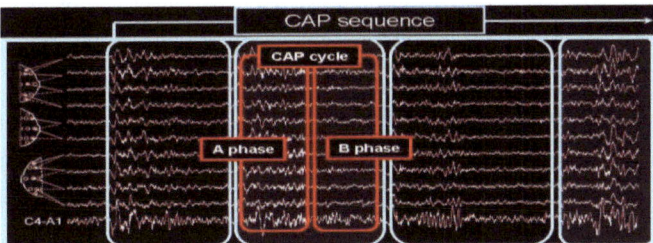

Figure 1.4: Cyclic Alternating Pattern

CAP sequences occur in all four stages (1, 2, 3 and 4) with usually 4 sleep onsets: in the period of after awakeness, during sleep and before the transition from NREM to REM sleep. During the normal REM sleep period CAP does not occur. The CAP sequence means that minimum of two CAP cycles must occur consecutively. The CAP cycle is composed of two phases A and B, when phase Are presents apparent changes in frequency and/or amplitude related with the training rhythm.

Therefore it is considered by cyclic arrangements of cerebral stimulation (phase A) tracked by epochs of deactivation which detached two successive phase A periods with an interval <1 min. The interval between two phases A is the phase B. Average duration of phase A is 10 – 12 seconds and of phase B is 20 – 30 seconds. Phase A periods are subdivided into three subtypes:

a) Subtype A1: synchronized events with low impact on autonomic and somatomotor activities;

b) Subtype A2: mixed synchronized–desynchronized EEG events with an intermediate influence on the autonomic and somato motor activities; and

c) Subtype A3: predominantly desynchronized EEG events with heavy effects on the autonomic and somato motor activities.

1.5 History

Many say the best way is to dismiss it from your mind and get on with life. One thing is for sure – insomnia is not going away any time soon. The beginning of contemporary studies of the sleep dates back to the mid-19th century. Before the understanding of the central nervous system, there was a couple of interesting theories about the cause of sleep. One of such theories considered the lack of blood, congestion or pressure of blood in the brain as the cause of sleep. Another theory provided an idea of the accumulation of toxic substances during the day, which simply caused sleep. Also the lack of oxygen was considered as a regular cause of every night sleep. Even after recognition of neurons and their electrical activity there was a long period of persuasion that neurons are paralyzed and activity is somehow switched off. In 1868, Wilhelm Griesinger described eyes movements during the sleep and referred them to as a part of dreaming and Sigmung Freud, in 1895, noticed the reduction of my tonus. Research that is more systematic was applied when the Austrian neurologist Constantine von Economy postulated the existence of an active sleep-regulating center in the brain and localized it. He described excessive sleepiness in patients with lesion in the back of the hypothalamus and insomnia in patients with lesion in the periotic area and in the front of the hypothalamus. The identification of such centers was based on clinical and path anatomical observations of patients with specific viral encephalitis, the lethargic encephalitis, in 1917 - 1920. Walter Rudolph Hess and Steven Walter Ranson later confirmed his findings. In 1920 Nathaniel Kleitman's research focused on how sleep and wakefulness relate to circadian rhythms and the effects of sleep deprivation. He decided on the cerebral cortex to be the site of wakefulness. Later in 1929 he proposed the theory that the inactivity and fatigue of the central nervous system and the loss of surrounding stimulation caused sleep. German scientist Hans Berger connects the modern sleep research with the invention of the electroencephalograph (EEG) in 1929. EEG is an instrument that measures and records brain wave patterns. In 1937, based on 30wholenight EEG measurements, the research group of Alfred Loomis discovered cycle repeating patterns and classified sleep into five different stages. In 1953 Nathaniel Kleitman and his student Eugene Aserinsky reported their findings about periods of eye movements during the sleep. They called that "Rapid eye movement sleep" (REM). Eye movements were rapid and binocularly symmetrical, EEG pattern was similar to one discovered in wakefulness and respiratory and heart rates were increased in contrast to other sleep periods. If the person was awoke during the period of REM sleep, he described highly vivid dreams, but if awoke during non-REM (NREM) period he did not recall dreams. Nathaniel Kleitman and William C. Dement decided on cyclic nature of sleep in 1955 and in the subsequent research recorded this cyclic pattern of REM and NREM sleep using the EEG and electrooculogram (EOG). From 4 to 5 cycles REM-NREM occur during the night while one cycle lasts around 90-100 minutes. Worldwide, many researches were focused on study of sleep disorders. One of the most cited author is Czech clinician Bedřich Roth, who helped to establish the first European sleep laboratory and directed the study at narcolepsy and hypersomnia. Allan Rechts chaffen and Anthony Kales focused their work on standardizationof scoring EEG recordings, so in 1968 their Manual of Standardized Terminology, Techniques and Scoring System for Sleep Stages of Human Subjects was published by the United States Government Printing Office. In 1992, the American Sleep Disorders Association (ASDA) formalized the rules for scoring central nervous system arousals and published the manual in the journal Sleep. Many researches are still based on this scoring manual in spite of the publication of the new one in 2007 by the American Academy of Sleep Medicine (AASM**).**

This new scoring manual invented by researchers of AASM contains guidelines of changed scoring rules. The Manual was published in a single volume entitled the AASM Manual for the Scoring of Snooze and Connected Proceedings: Rules, Lexicon and Technical Specifications and represents a giant step towards standardization of clinical polysomnography. Even as scoring rules were evolving with changing methodology, no official guidelines emerged concerning computerized polysomnography.

1.6 Motivation

Insomnia is a prevalent health complaint, and the incidence upsurges with mature. Methods such as reduction exercises snooze restriction treatment and overhauling may be valuable. Hence, its analysis and action is a requirement.

CHAPTER 2
LITERATURE REVIEW

2.1 Introduction

Insomnia is a collective and extensive complaint with a significant impact on both night-time and daytime functioning. Insomnia has been mainly explained by behavioral and neurocognitive models. Especially in the last two decades, The quantity of sluggish wave action in non-rapid eye movement (NREM) snooze is reflected a marker of NREM sleep intensity and the electrophysiological correlate of a sleep-wake dependent 'Process S' underlying sleep homeostasis, a process influenced by different physiological[16] or experimental conditions.

Many investigations carried out using quantitative analysis support the hypothesis that primary insomnia is associated with hyperarousal of central nervous system (CNS), because patients with insomnia (PI) exhibited increased high frequency EEG activity during both sleep onset and all-night sleep. Some studies support the hypothesis that sleep homeostasis is altered in primary insomnia, as expressed by a slow wave sleep (SWS) deficiency, and that homeostatic deregulations may represent a predisposing, precipitating and/or perpetuating factor of insomnia. Recently, it was also suggested that hyper arousal and altered sleep homeostasis (and even circadian deregulation) may interact simultaneously in chronic insomnia [18].

Nevertheless, the body of research that provided these results has some intrinsic limits, often attributed to standard taxation and clinical investigation methods.

a) Between subjects designs seem to be inadequate to assume generalizable features in patients with snooze disorders. Evidence from basic sleep research suggests that normal snooze is considered by large separate changes, which could constitute a confounding factor in the evaluation of the physiological basis in pathological sleep. A growing body of evidence points to genetic influences on normal and pathological sleep, in humans and in animals. As an example, it has been shown that a stable frequency-specific (8.0-15.5 Hz) decoration of EEG structure laterally the antero-posterior cortical axis during NREM sleep distinguishes each individual like a "fingerprint", reflecting genetic influences. Therefore, further studies with larger samples of patients and within-subjects designs (i.e., longitudinal studies) are needed.

b) Numerous trainings on PI recycled EEG data resulting from the leading in-laboratory education night. One single night of recording appears to be insufficient to assess the survival of steady EEG alterations and the 'first-night effect' might interfere with results.

c) Some studies were performed on not completely drug-free PI, making it difficult to discriminate the exact pharmacological influence on rapid occurrence EEG bands.Lastly, we propose views for future research, by an exemplification of integration of knowledge from basic and clinical research.

2.2 Degree of pineal calcification (DOC) is associated with polysomnographic sleep measures in primary insomnia patients

According to this paper, melatonin shows a important role in the suitable operative of the daily timing arrangement & exogenous melatonin is helpful in the CTS & sleep disorders. Limit the connection between the grade of pineal calcification & a variety of sleep limitations measured accurately using PSG. The total no of patient is 31 with main insomnia were comprised in our education. An alteration night, polysomnography footage was achieved in the sleep workroom. Urine samples were composed at 32-h period included the polysomnography night. The amount of 6-sulphatoxymelatonin stages was resolute using ELISA. Degree of pineal calcification and capacity of calcified pineal material & unclarified pineal matter were probable through the cranial computed tomography.

2.3 Studying Herb-Herb Interaction for Insomnia through the theory of Complementarities

According to this paper, the value of a TCM medicine arises from the herb herb interface in a method. It is not relaxed to define the accurate interrelating herbs to donate to the efficacy of a cure. Implication rule removal is a method to find the co incidence of some matters. It is not areas concerned with the produced consequences are very delicate to the given limitations such as support amount. The aim of this title is to familiarize a new outline too methodically produce a set of groupings of relating herbs that tips to good conclusion.

2.4 Internet-delivered or mailed self-help treatment for insomnia? A randomized waiting-list controlled trial

According to this paper, the effects of the help of CBT have been unpredictable. The purpose of this learning was to define the success of self-help for insomnia distributed in any electronic format associated to a waiting-list. Members kept a record and occupied out surveys earlier to the randomized into n ¼ 216, n ¼ 205 groups. The involvement contained of six weeks of unfounded self-help CBT & post-tests were 18 & 4 weeks after interference. At four week follow up electronic and paper-&-pencil circumstances were superior through the waiting-list disorder on most everyday sleep events, worldwide insomnia symptoms depression & anxiety indications. The electronic & paper-&-pencil collections established equal efficacy four weeks after cure belongings were continued at forty eight week continuation.

2.5 Insomnia Characterization: From Hypnogram to Graph Spectral Theory

According to this paper, calculate & differentiate control & insomnia snooze onset arrangements by biomedical signal treating of instant PSG. The method contained of three tandem elements like bio signal treating module which used state space time fluctuating autoregressive touching average methods with recursive spot filter, hypnogram group module that applied a fuzzy implication system, insomnia characterization element that separated between mechanism & topics with insomnia by a logistic reversion model qualified with a usual of resemblance events. The education employed sleep arrival phases from sixteen control & sixteen topics with insomnia. FIS achieved automatic sleep scoring vigorous to inter-subjects & inter-raters erraticism. A geometric usual of unpaired two-tailed t-examinations suggested that expanses d1, d2 & d3 had greater statistical consequence to characterize sleeping decorations. The overview of graph phantom theory & logistic reversion for the analysis of insomnia signifies a novel idea.

2.6 Relation Between Heart Beat Fluctuations and Cyclic Alternating Pattern During Sleep in Insomnia Patients

According to this paper, 30 % of the populace among 18 years and 60 years grieves from insomnia. The special effects of this illness involve difficulties such as deprived school or job recital and transportation accidents. In addition, patients with insomnia current deviations in the cardiac purpose throughout sleep. Besides, the edifice of electroencephalographic

A-phases, which shapes up the Recurring Flashing Design throughout sleep, is associated to the insomnia trials. The connection between these mind activations and the autonomic worried system would be of attention, tightfitting the interaction of essential and autonomic movement throughout insomnia with this goal, a study of the relationship between A-phases and temperament rate variations is accessible. Polysomnography footage of five strong subjects, five sleep misperception patients & five patients through psychophysiological insomnia were used in the study. DE trended Fluctuation Analysis was recycled in directive to assess the heart rate undercurrents and this was connected with the amount of A-phases. The effects recommend that compulsive patient's current changes in the subtleties of the sentiment degree.

2.7 Processing of Signals Recorded Through Smart Devices: Sleep-Quality Assessment

According to this paper, we revenue into reflection the emotion rate erraticism and breathing signals for involuntary sleep performance, stimulations detection, and apnea gratitude. This is mainly suitable for wearable or material devices that might be employed for family monitoring of sleep. The heart rate variability and the breathing were analyzed in the incidence domain, and the indicators on the spectral and cross-spectral limitations put into indication the prospect of a sleep assessment on their origin. Evaluation with customary polysomnography exposed a classification correctness of 89.9% in rapid eye effort and non-rapid eye effort snooze withdrawal and an exactness of 88% for sleep apnea exposure. Further information can be realized from the quantity of micro arousals familiar in communication of typical amendments in the heart rate variability signal. The acquired results funding the idea of automatic sleep evaluation and watching over signals that are not usually used in irrefutable polysomnography, but can be simply recorded at home concluded wearable devices.

2.8 Advantages and Challenges of Power Spectral Density Estimation Methods for Scanning Radar Angular Super resolution

According to this paper there are many techniques documented in literature to enhance the angular resolution, of which DE convolution method and power spectral we focus on analyzing the advantages and challenges of PSD methods in comparison with the deconvolution method. Firstly, three typical PSD estimation approaches are introduced, followed with the comparison with deconvolution density methods are favored and attain many interests. In this paper, method that summarizes the advantages and challenges of Power Spectral Density methods in theory. Simulations are provided in terms of coherence and number of snapshots, which presents the performance of different PSD methods and Lucy-Richardson de convolution method, healthier representative the compensations and tests of PSD procedures.

2.9 Modelling and Control for Simultaneous Laser Beam Alignment of a Dual-PSD Industrial Robot Calibration System

According to this paper,a critically important role for improving the accuracy of industrial robots. Recently, position sensitive device based calibration methods provide a more efficient way to perform joint offset calibration than the conventionally used methods. We have proposed a dual-PSD based method to accurately calibrate the joint offsets of industrial robots. However, the required alignment of laser beam to shoot at the two PSD centers was performed separately, which required to pre-know some parameters for the alignment task. This paper proposes a simultaneous laser beam alignment method without knowing these parameters to perform the laser beam alignment task. The modeling and control for the alignment systems are derived and simulation results exemplify the efficiency of the recommended methods.

2.10 Deep Learning EEG Response Representation for Brain Computer Interface

According to this paper, the multi-scale deep convolutional neural networks are introduced to deal with the representation for imagined motor Electroencephalography (EEG) signals. We propose to learn a set of high-level feature representations through deep learning algorithm, referred to as Deep Motor Features for brain computer interface with imagined motor tasks. As the extracted Deep Motor Features are dissimilar for different tasks and alike for the same tasks, it is convenient to separate the miscellaneous EEG signals for fantasy motor tasks separately. Our approach achieves 100% correctness for 4 modules imagined mechanical EEG signals ordering on Project BCI - EEG motor activity dataset. Moreover, thanks to the highly abstract features Deep Motor Features learned, only 4.125 seconds trials of training data are needed, compared with the conventional BLDA algorithm for 8.75 seconds trials demand to achieve the same accuracy, accordingly the BCI response time and the required trials for training are almost declined by half. Experiments are provided to illustrate the effectiveness of the proposed design approach.

2.11 A New Algorithm for Noise PSD Matrix Estimation in Multi-Microphone Speech Enhancement Based on Recursive Smoothing

According to this paper, we present a new algorithm for the estimation of the noise power spectral density matrix, as needed for multi-microphone speech enhancement in a general non-stationary noisy environment. First, we propose a recursive scheme for noise PSD estimation in which the current, previous and close subsequent noisy speech frames are properly weighted. The forgetting factor for the recursive updating of the smoothed PSD is obtained based on an overall measure of the SNR across all microphone signals. Since this SNR measure depends on the noise statistics, we choose to iteratively update it using the latest available estimate of the noise PSD matrix. Finally, to obtain better estimation accuracy in the proposed method, we further apply a direct extension of the minimum tracking approach to the estimated noise PSD matrix. Performance of the proposed algorithm is evaluated in terms of objective measures and its superiority is shown with respect to two recent noise PSD estimation methods in the situation of speech improvement.

2.12 Power Spectral Density and High Order Bi-spectral Analysis of Alzheimer's EEG

According to this paper, power spectrum density and high order bi-spectral analysis are proposed to investigate the abnormalities of electroencephalogram signals which are collected from fourteen Alzheimer's disease patients and fourteen age-matched normal controls. The power spectrum density estimated by AR model is first used to evaluate the frequency distribution in AD brain. It is found that for the whole frequency band, the significant alteration occurs at the electrode FP1, FP2, O1 and T3, and for the sub-bands, the relative PSD shows significant increase in the theta incidence band diminution in the alpha 2 incidence bands in AD. In order to explore the nonlinear phase-coupling information, high order bispectral analysis is further applied to O1 electrode in the whole frequency band. It is demonstrated that the non gaussianity and nonlinearity of EEG are decreased in AD by bispectral analysis. The obtained results show that analysis of PSD and BIS can be taken as a potential comprehensive measure to distinguish AD patients from the standard, which may advantage our sympathetic of the syndrome.

2.13 Binaural Noise Suppression based on an unbiased estimator of target PSD in Complex Noise Environments

According to this, paper a biased target PSD because of irrelevant assumptions for the noisy atmosphere. In this tabloid, an impartial target power spectral density is obtained by removing the effect of wordy noise on the anticipation strainer. In addition, by on-line estimation of both the noise power spectral density and target transfer function ratio from the input signals, the proposed algorithm achieves robust noise suppression for an unidentified target route under a reckless time-varying noisy environment.

2.14 Detection of rapid eye movement behaviour disorder using short time frequency analysis of PSD approach applied on EEG signal (ROC-LOC)

According to this paper, snooze syndrome is a remedial sickness of the sleep patterns of a creature. Some snooze syndromes are somber adequate to barricade with normal significant, mental and demonstrative routine. Quality and waveform of electroencephalogram indications of human existence examined. This paper is to attraction the consequence in the procedure of gesture spectrum examination of the variations in the territory of different periods of catnap.

CHAPTER 3
RESEARCH METHODOLOGY

This chapter deals with the methods used to test the hypothesis that sleep disorder events can be detected by changes in the power spectral density that occur due to the termination of these events by cortical arousals.

3.1 Characteristics of the EEG signal

Various methods that have been tried to extract quantitative features from an EEG signal have always faced tasks payable to the detail that the subtleties of EEG depends on brain activities which in turn are related to processing of information that originates internally as well as externally. It is revealed that the EEG signal was a highly non stationary process and it only could be described by the basic stochastic concepts for durations not longer than 10-20s. A study showed that the variability of power of the main spectral EEG components for segments between 5 to 10 s ranged up to 50-100%. Hence they concluded that in order to determine its spectrum the signal should be analyzed as a series of stationary random processes. Such processes have average values that are constant and autocorrelation purposes that be contingent single on time changes. Such signals have finite average power and hence are considered by a PSD. In repetition, a single gratitude of the random process is considered and an estimate of the power spectrum of the process is computed.

3.2 Experimental Setup

Each subject was tested for one night for approximately 8 hours. The standard polysomnographic (NPSG) data, which included ECG, electroencephalogram (EEG), EOG and electromyogram were logged on the data acquisition computer. From that we took only the EEG data for one minute.

3.2.1 Subject Details

Total nine subjects were taken who were suffering from the sleep disorder of insomnia. The subject's details like gender, age, sleep duration of each stage etc (table 3.1). The total Normal Patient is sixteen all data of the normal patient is given the table 3.2

Table 3.1: Insomnia Patients Details of duration of start time and end time of 1 minute

Sl No.	Patient Details			Sleep Time Duration of S_0 Sleep Stage		Sleep Time Duration of S_1 Sleep Stage		Sleep Time Duration of S_2 Sleep Stage		Sleep Time Duration of S_3 Sleep Stage		Sleep Time Duration of S_4 Sleep Stage		Sleep Time Duration of REM Sleep Stage	
	ID	GEN	AGE (Years)	Start Time	End Time	Start Time	End Time	Start Time	End Time	Start Time	End Time	Start Time	End Time	Start Time	End Time
1	INS 1	M	54	22:32:28	22:33:28	22:57:58	22:58:58	23:09:58	23:10:58	23:39:28	23:40:28	N/A	N/A	04:04:58	04:05:58
2	INS 2	F	58	18:26:38	18:27:38	N/A	N/A	23:24:08	23:25:08	00:16:38	00:17:38	N/A	N/A	02:31:38	02:32:38
3	INS 3	M	82	23:18:42	23:19:42	01:49:12	01:50:12	22:12:42	22:13:42	02:51:42	02:52:42	03:09:12	03:10:12	04:49:12	04:50:12
4	INS 4	F	58	21:45:34	21:46:34	N/A	N/A	21:53:04	21:54:04	22:36:34	22:37:34	00:51:34	00:52:34	22:54:34	22:55:34
5	INS 5	F	59	17:58:48	17:59:48	23:30:18	23:31:18	23:32:48	23:33:48	01:54:18	01:55:18	02:03:48	02:04:48	03:11:48	03:12:48
6	INS 6	F	54	22:37:17	22:38:17	00:54:17	00:55:17	00:00:17	00:01:17	01:35:47	01:36:47	00:23:17	00:24:17	03:36:47	03:37:47
7	INS 7	F	47	20:28:14	20:29:14	05:51:14	05:52:14	22:39:14	22:40:14	23:55:14	23:56:14	22:18:14	22:19:14	23:24:14	23:25:14
8	INS 8	M	64	22:43:04	22:44:04	23:39:04	23:40:04	23:18:34	23:19:34	00:43:34	00:44:34	N/A	N/A	04:06:34	04:07:34
9	INS 9	M	72	04:07:44	04:08:44	05:05:14	05:06:14	23:23:14	23:24:14	23:40:14	23:41:14	02:38:44	02:39:44	02:40:14	02:41:14

Table 3.2: Normal Patients Details of duration of start time and end time of 1 minute

Sl No.	Patient Details			Duration of S_0 Sleep Stage		Duration of S_1 Sleep Stage		Duration of S_2 Sleep Stage		Duration of S_3 Sleep Stage		Duration of S_4 Sleep Stage		Duration of REM Sleep Stage	
	ID	GEN	AGE (Years)	Start Time	End Time	Start Time	End Time	Start Time	End Time	Start Time	End Time	Start Time	End Time	Start Time	End Time
1	N1	F	37	22:11:03	22:12:03	06:19:03	06:20:03	23:29:33	23:30:33	0:35:03	0:36:03	22:40:33	22:41:33	23:36:03	23:37:03
2	N2	M	34	22:21:06	22:22:06	00:44:06	00:45:06	23:20:06	23:21:06	22:47:06	22:48:06	22:52:06	22:53:06	05:00:36	05:01:36
3	N3	F	35	23:11:42	23:12:42	03:02:42	03:03:42	23:55:12	23:56:12	23:20:42	23:21:42	23:25:12	23:26:12	03:20:12	03:21:12
4	N4	F	25	22:39:37	22:40:37	05:34:37	05:35:37	23:54:07	23:55:07	3:33:07	3:34:07	0:29:07	0:30:07	04:09:07	04:10:07
5	N5	F	35	22:50:18	22:51:18	22:53:18	22:54:18	0:01:18	0:02:18	0:52:48	0:53:48	1:05:48	1:06:48	04:35:18	04:36:18
6	N6	M	31	NA	NA	NA	NA	NA	NA	NA	NA	NA	NA	NA	NA
7	N7	M	31	NA	NA	NA	NA	NA	NA	NA	NA	NA	NA	NA	NA
8	N8	F	42	22:21:11	22:22:11	06:27:41	06:28:41	6:02:11	6:03:11	22:56:41	22:57:41	23:00:41	23:01:41	04:44:11	04:45:11
9	N9	M	31	NA	NA	NA	NA	NA	NA	NA	NA	NA	NA	NA	NA
10	N10	M	23	23:27:22	23:28:22	NA	NA	1:07:22	1:08:22	1:34:22	1:35:22	1:41:22	1:42:22	04:00:52	04:01:52
11	N11	F	28	22:40:46	22:41:46	NA	NA	23:08:46	23:09:46	0:38:16	0:39:16	23:36:46	23:37:46	03:32:16	03:33:16
12	N12	M	29	15:16:22	15:17:22	15:29:22	15:30:22	15:30:52	15:31:52	15:38:52	15:39:52	15:55:22	15:56:22	20:03:52	20:04:52
13	N13	F	24	**	**	**	**	**	**	**	**	**	**	**	**
14	N14	F	35	**	**	**	**	**	**	**	**	**	**	**	**
15	N15	M	34	22:01:22	22:02:22	22:16:52	22:17:52	22:28:52	22:29:52	22:34:52	22:35:52	23:36:52	23:37:52	03:01:52	03:02:52
16	N16	F	41	22:37:47	22:38:47	06:17:47	06:18:47	23:58:17	23:59:18	6:54:47	6:55:47	23:12:47	23:13:47	06:04:47	06:05:47

3.3 Algorithm

Step 1: EEG Signal Extraction

The study was based on data that was collected and processed by the physionet.org. Physiobank collections are organized into more than 50 databases, each containing a number of records, and each record containing information collected from a single subject. We have downloaded one minute data on insomnia in all the sleep stages and cut the selected channels: EMG1-EMG2, FP2-F4

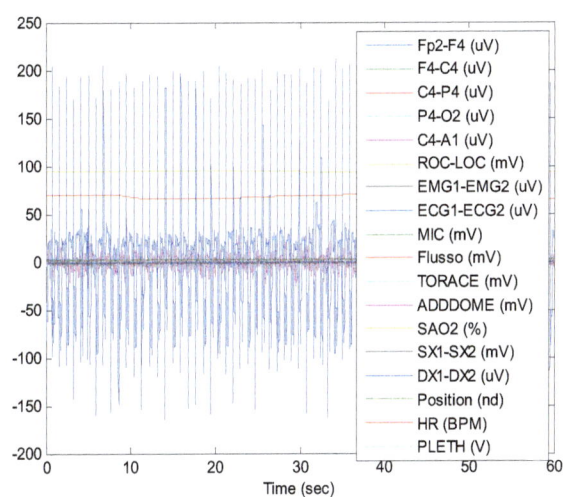

Figure 3.1: Complete EEG Signal

Step 2: Removal of DC Component

In a majority of the EEG examinations, scalp electrodes that are used are not in direct contact with the tissue. The electrolyte bridge that is formed by an electrode jelly that is applied between the electrode and the skin establishes an indirect contact. A steady potential (DC offset voltage) is created at this junction depending on the electrolyte composition and the condition of the skin which can be as large as the magnitude of the electrical activity recorded from the brain. The DC component was removed from the clipped signal by removing the best straight line fit from the data. The first component of signal's FFT is actually the mean of complete signal and it signifies signal largeness at 0 incidence, so by making it to zero.

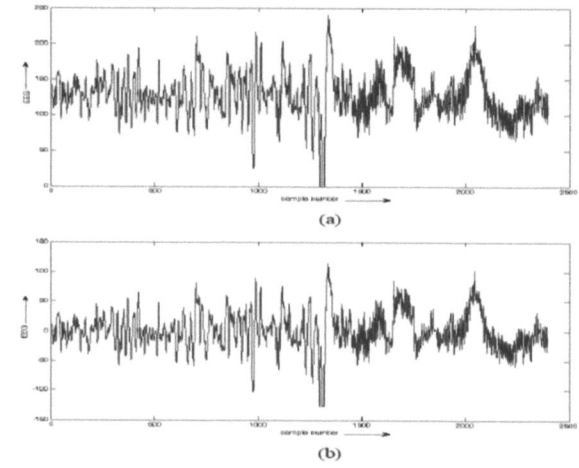

Figure 3.2: EEG clip data having DC offset (a) and having DC offset removed (b)

Step 3: Removal of high frequency components

The DC offset free EEG signal is then filter dosing Low pass FIR(Finite Impulse Response) filter of cut off frequency 25 Hz. FIR filters are designed using the Signal Processing functions and Direct-Form II Transpose Filter. An FIR filter of order 200 is designed using a Hanning window. Now filter function 'filtfilt' performs zero-phase digital filtering by dispensation the input documents, in equally the forward and reverse direction. After filtering the data in the forward direction, 'filtfilt' converses the drinkable order and turns it back through the filter. The result has the following characteristics:

a) Zero-phase distortion

b) A filter transfer function, which equals the squared magnitude of the original filter transfer function

For FIR filter, the syntax is:

b = fir1(n,Wn,window)

Where, Wn is a number between 0 & 1, and 1 corresponds to the Nyquist frequency.

For filtered signal, the syntax is:

y = filtfilt (b,a,x)

Where vector b provides the numerator constants of the strainer and the path a delivers the denominator coefficients and x is the input data.

Step 4: Calculation of N point FFT for Power Spectral Density

Suppose that the signal representation in the frequency domain is X(ω) and is periodic with a period 2π, only samples in the fundamental frequency range are necessary to compute. Considering the incidence domain is appraised at N intermediate samples from 0 to fs, and then the interval between two successive samples is δf. This is written mathematically as:

δf =f*n/fs

N=8192 spectral samples, fs=512 Hz

Assuming δf=16 sample per Hz change.

We consider N=8192(next power of 2)

Step 5: PSD Estimator (Welch Method)

PSE is most important application area in Digital Signal Processing. There are mainly two types of power spectrum estimation (PSE) method: Parametric and nonparametric.

Welch method is nonparametric method that include the periodogram that have the advantage of possible implementation using the fast Fourier Transform. The periodogram is a method of estimating the autocorrelation of finite length of a signal. The periodogram technique based on Welch method is capable of providing good resolution if data length samples are selected optimally. It can be observed that PSE based on the Hamming give better results than Hanning window.

$$P_{xx}(f) = \frac{1}{LF_s} | \sum_{n=0}^{L-1} x_L(n) e^{-j2\pi fn/F_s} |^2$$

Where F_s is the sampling frequency.

The actual computation of $P_{xx}(f)$ can be performed only at a finite number of frequency points, and usually employs an FFT. Most executions of the periodogram technique calculate the N-point PSD estimate at the frequencies:

$$f_k = \frac{kF_s}{N} \quad k = 0,1,...,N-1$$

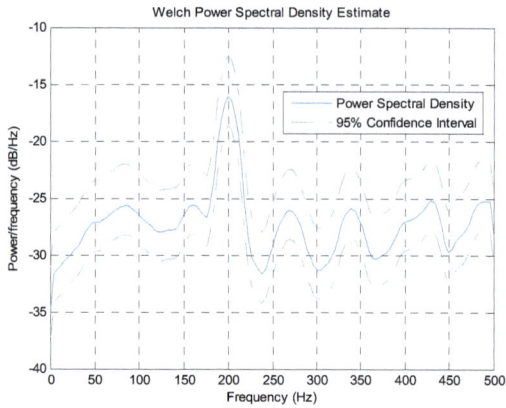

Figure 3.3: Welch Power Spectral Density Estimate

Step 6: Area calculation by using trapezoidal method

Area Estimation of delta, theta, alpha, gamma frequency bands are calculated by using Trapezoidal method. Delta (δ) wave having frequency range 0.5 to 4 Hz, theta (θ) wave having frequency range 4 to 8Hz, alpha (α) wave having frequency range 8 to 13 Hz, beta (β) wave having frequency range 13 to 30 Hz.

Step 7: Power ratios are calculated by dividing average power of individual wave frequency by total average power of all bands

In order to calculate the average power for individual frequency bands for the entire night, initial steps of DC component removal and low pass filtering were performed. The average power in a given frequency band technique usages a rectangle calculation of the fundamental of the signal's power spectral density (PSD). The average power is the total signal power and the total average power is contained in the one-sided or two-sided spectrum. The PSD estimate was calculated and average power of frequency bands delta, theta, alpha and beta were calculated and then normalized.

CHAPTER 4
RESULTS AND DISCUSSIONS

4.1 Loading the signal

Figure 4.1 shows complete EEG signal of a single stage. On giving command load (matname) the signal with matname 'ins1_edfm.mat' is loaded in MATLAB workplace and the title of numerous signal and their details are loaded from file 'ins1_edfm.info'.Load (matname) knack stretches a signal in workplace named as 'val'.This figure shows the full signals and is on time basis. The signal is of one minute.

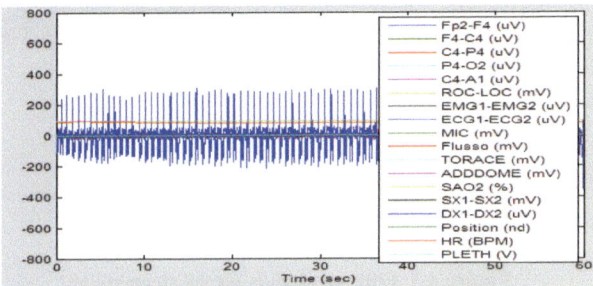

Figure 4.1: Complete EEG Signal of S0 Stage

4.2 Extracting EEG Signals

Now the second step is to separate the channel from the extracted EEG signal. This helps in studying waveform of each channel separately. In this work, two channels have been considered for study. They are mentioned below:

EMG1-EMG2, FP2-F4

All stages have been considered for the above mentioned channels. These stages are S0, S1 and REM.

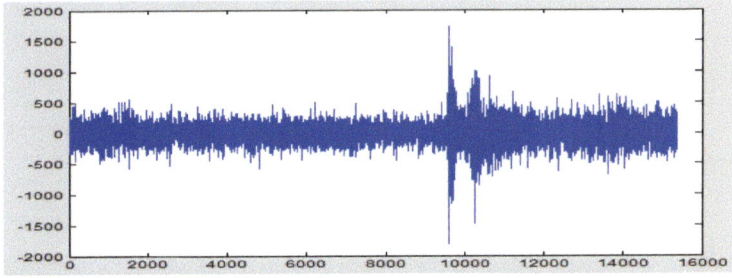

Figure 4.2: An extracted signal with EMG1-EMG2

4.3 Filtering of EEG Signals

In this step we basically label the X-axis. Time in second is labeled on the X-axis. This figure depicts the signal that is filtered using low pass filter. A low pass filter of cut off frequency 25 Hz is used.

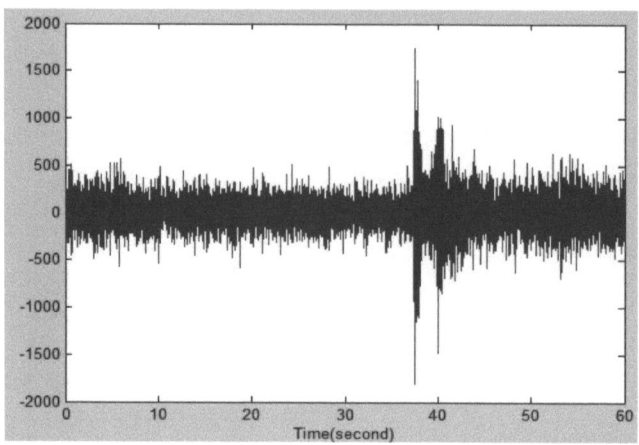

Figure 4.3: Filtered signals for S0 sleep stage

4.4 Applied Window

The next figure shows the output of the filter that has been obtained by first inverting the output of the filter then passing it again through the filter to obtain a response with phase band cancellations present in the previous output.

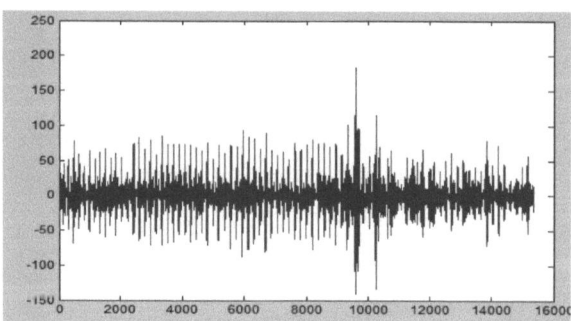

Figure 4.4: Window signals for S_0 Sleep Stage

4.5 Comparison of the Original Wave & Filtered Wave

The differences between original and filtered waveform is shown in figure 4.5. The MATLAB function 'filtfilt', which is a zero-phase filtering, is used as the cleaning technique. Original and filtered waveform is cut in 1:1000 ratios which gives the detail on the minute differences of both signals.

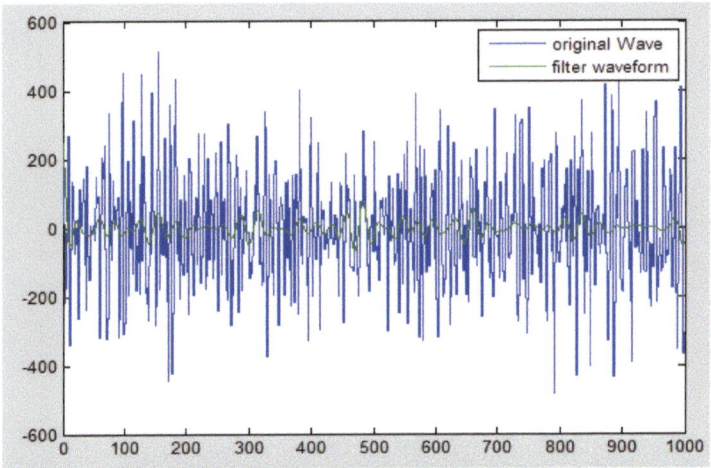

Figure 4.5: Differences between original and filtered waveform

4.6 Estimation of power spectral density

An improved estimator of the PSD is the one proposed by Welch. The method consists of distributing time series statistics into sections, computing a modified periodogram of each segment, and then averaging the PSD approximations. The outcome is Welch's power spectral density estimation. Welch's method is implemented in the toolbox by the 'spectrum. welch' object. Since PSD gives signal power with respect to the frequency spectrum, we require specifying the number of incidence slots to allocate the spectral command. It is called as number of FFT points. The figure 4.6 is showing power spectral density of different channels. This figure shows the power spectral density estimate done using Welch method of the power spectral density estimation.

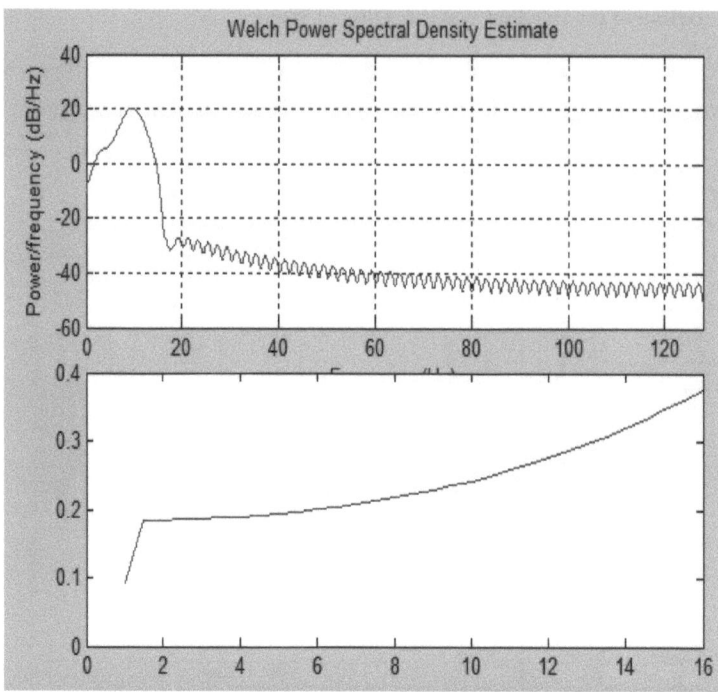

Figure 4.6: Estimated PSD of different channels

4.7 Calculation of normalized power

After taking PSD of S0, we have calculated average power for EEG waves from power estimation. For this purpose we applied the finite integral of the PSD estimate over the limit related to the frequency range of particular wave.

Following tables of FP2-F4 channel of S0, S1 and REM sleep stages is given below which shows normalized power of both insomnia patient and normal patient.

CHAPTER 5
CONCLUSION

Normalized power (Pnorm) of normal cases having no symptoms of sleep is analyzed and compared with pathological cases during S0 stage. Normalized power indicates the percentage of a particular EEG activity out of complete power. So it is found that it gives a better indication of measurements of detection of features instead of taking average power of particular EEG activity. We took 16 normal cases and 9 insomnia cases and then compared the results.

Table 5.1: Normalized Power of the theta Wave of Normal Patient and Insomnia Patient for FP4-F2 Channel and Stage S1

StageS$_1$	N2	N3	N11	Ins 3	Ins 5	Ins 6
P_Theta	0.37561	0.31792	0.28589	0.25685	0.26282	0.1876
Remark	High			Low		

The above table shows the normalized power of Theta wave for Normal patient and the Insomnia patient. The study is done for FP4-F2 channel and stage S1. From the table, it is observed that normalized power of Theta wave for Normal patient lie between 0.28589 and 0.37561 While those of Insomnia patient is between 0.1876 and 0.26282. Thus, we can say normalized power of Theta wave in case of normal patient is high and that of Insomnia patient is low.

Table 5.2: Normalized Power of the alpha Wave of Normal Patient and Insomnia Patient for FP4-F2 Channel and Stage S1

StageS$_1$	N1	N2	N5	Ins 3	Ins 5	Ins 7
P_Alpha	0.072045	0.10871	0.054428	0.15832	0.16001	0.11647
Remark	Low			High		

The above table shows the normalized power of Alpha wave for Normal patient and the Insomnia patient. The study is done for FP4-F2 channel and stage S1. From the table, it is observed that normalized power of Alpha wave for Normal patient lie between 0.054428and 0.10871. While those of Insomnia patient is between 0.11647 and 0.16001. Thus, we can say normalized power of Alpha wave in case of normal patient is low and that of Insomnia patient is high.

Table 5.3: Normalized Power of the theta Wave of Normal Patient and Insomnia Patient for EMG1-EMG2 Channel and Stage S_0

StageS_0	N1	N5	N11	Ins I2	Ins 4	Ins 7
P_Theta	0.27951	0.29706	0.2684	0.32262	0.3637	0.3034
Remark	Low			High		

The above table shows the normalized power of Theta wave for Normal patient and the Insomnia patient. The study is done for EMG1-EMG2 channel and stage S0. From the table, it is observed that normalized power of Theta wave for Normal patient lie between 0.2684 and 0.29706. While those of Insomnia patient is between 0.3034 and 0.3637. Thus, we can say normalized power of Theta wave in case of normal patient is low and that of Insomnia patient is high.

Table 5.4: Normalized Power of the alpha Wave of Normal Patient and Insomnia Patient for EMG1-EMG2 Channel and Stage S_0

StageS_0	N1	N2	N5	Ins 1	Ins 3	Ins 6
P_Alpha	0.56302	0.55817	0.57128	0.71758	0.72784	0.69462
Remark	Low			High		

The above table shows the normalized power of Alpha wave for Normal patient and the Insomnia patient. The study is done for EMG1-EMG2 channel and stage S0. From the table, it is observed that normalized power of Alpha wave for Normal patient lie between 0.55817 and 0.57128. While those of Insomnia patient is between 0.69462 and 0.72784. Thus, we can say normalized power of Alpha wave in case of normal patient is low and that of Insomnia patient is high.

Table 5.5: Normalized Power of the theta Wave of Normal Patient and Insomnia Patient for EMG1-EMG2 Channel and Stage S_0

Stage S_0	N1	N5	N11	Ins 2	Ins 4	Ins 5
P_Beta	0.10351	0.1191	0.11575	0.019096	0.01633	0.011432
Remark	High			Low		

The above table shows the normalized power of Beta wave for Normal patient and the Insomnia patient. The study is done for EMG1-EMG2 channel and stage S0. From the table, it is observed that normalized power of Beta wave for Normal patient lie between 0.10351 and 0.1191. While those of Insomnia patient is between 0.011432 and 0.01633. Thus, we can say normalized power of Beta wave in case of normal patient is high and that of Insomnia patient is low.

Table 5.6: Normalized Power of the theta Wave of Normal Patient and Insomnia Patient for EMG1-EMG2 Channel and Stage S_1

Stage S_1	N 3	N 5	N 11	Ins 5	Ins 6	Ins 9
P_Theta	0.31338	0.31384	0.31897	0.42748	0.32362	0.34105
Remark	Low			High		

The above table shows the normalized power of Theta wave for Normal patient and the Insomnia patient. The study is done for EMG1-EMG2 channel and stage S1. From the table, it is observed that normalized power of Theta wave for Normal patient lie between 0.31338 and 0.31897. While those of Insomnia patient is between 0.34105 and 0.42748. Thus, we can say normalized power of Theta wave in case of normal patient is low and that of Insomnia patient is high.

Table 5.7: Normalized Power of the alpha Wave of Normal Patient and Insomnia Patient for EMG1-EMG2 Channel and Stage S_1

Stage S_1	N 2	N 11	N 16	Ins 3	Ins 7	Ins 9
P_Alpha	0.47316	0.54188	0.12052	0.69549	0.62897	0.5431
Remark	Low			High		

The above table shows the normalized power of Alpha wave for Normal patient and the Insomnia patient. The study is done for EMG1-EMG2 channel and stage S1. From the table, it is observed that normalized power of Alpha wave for Normal patient lie between 0.12052 and 0.54188. While those of Insomnia patient is between 0.5431 and 0.69549. Thus, we can say normalized power of Alpha wave in case of normal patient is low and that of Insomnia patient is high.

Table 5.8: Normalized Power of the beta Wave of Normal Patient and Insomnia Patient for EMG1-EMG2 Channel and Stage S_1

Stage S_1	N 3	N 5	N 11	Ins 6	Ins 7	Ins 9
P_Beta	0.084361	0.087815	0.083583	0.016274	0.018783	0.019323
Remark	High			Low		

The above table shows the normalized power of Beta wave for Normal patient and the Insomnia patient. The study is done for EMG1-EMG2 channel and stage S1. From the table, it is observed that normalized power of Beta wave for Normal patient lie between 0.083583 and 0.87815. While those of Insomnia patient is between 0.016274 and 0.019323. Thus, we can say normalized power of Beta wave in case of normal patient is high and that of Insomnia patient is low.

Table 5.9: Normalized Power of the theta Wave of Normal Patient and Insomnia Patient for EMG1-EMG2 Channel and Stage REM

StageREM	N 3	N 6	N 10	Ins 2	Ins 4	Ins 5
P_Theta	0.35896	0.35398	0.35682	0.47154	0.46438	0.45064
Remark	Low			High		

The above table shows the normalized power of Theta wave for Normal patient and the Insomnia patient. The study is done for EMG1-EMG2 channel and stage REM. From the table, it is observed that normalized power of Theta wave for Normal patient lie between 0.35398 and 0.35896. While those of Insomnia patient is between 0.45064 and 0.47154. Thus, we can say normalized power of Theta wave in case of normal patient is low and that of Insomnia patient is high.

Table 5.10: Normalized Power of the alpha Wave of Normal Patient and Insomnia Patient for EMG1-EMG2 Channel and Stage REM

Stage REM	N 3	N 5	N 7	Ins 5	Ins 7	Ins 8
P_Alpha	0.50534	0.50999	0.50389	0.34491	0.32946	0.38143
Remark	High			Low		

The above table shows the normalized power of Alpha wave for Normal patient and the Insomnia patient. The study is done for EMG1-EMG2 channel and stage REM. From the table, it is observed that normalized power of Alpha wave for Normal patient lie between 0.50389 and 0.50999. While those of Insomnia patient is between 0.32946 and 0.38143. Thus, we can say normalized power of Alpha wave in case of normal patient is high and that of Insomnia patient is low.

Table 5.11: Normalized Power of the beta Wave of Normal Patient and Insomnia Patient for EMG1-EMG2 Channel and Stage REM

Stage REM	N 1	N 5	N 10	Ins 2	Ins 4	Ins 7
P_Beta	0.050419	0.050891	0.055018	0.0066997	0.0093387	0.0049383
Remark	High			Low		

The above table shows the normalized power of Beta wave for Normal patient and the Insomnia patient. The study is done for EMG1-EMG2 channel and stage REM. From the table, it is observed that normalized power of Beta wave for Normal patient lie between 0.50419 and 0.55018. While those of Insomnia patient is between 0.0049383 and 0.0093387. Thus, we can say normalized power of Beta wave in case of normal patient is high and that of Insomnia patient is low.

From the above tables we have concluded that for the channel EMG1-EMG2 and by considering the theta waves the normalized power of the wave is low and Insomnia patient is high for s0,s1 and REM stage of sleep. Now by considering the alpha wave the normalized power of normal patient is low and high for insomnia patient for s0 and s1 stage of sleep, whereas high for normal in REM stage and low for insomnia patient. In beta wave the normalized power of s0, s1 and REM stage of sleep is high for normal patient and low for insomnia patient.

FUTURE SCOPE

In this research work, the study of rapid eye movement sleep behavior disorder using short-time frequency analysis applied on EEG signal is done. Here, different stages of sleep, sleep disorders, and Insomnia disorder have been discussed in detail. Rapid eye movement disorder is just like other diseases that need to be taken care of. But most of us are still not conscious about the severity of the diseases. As earlier the methods were graphical so to diagnose them was a big issue. This method will not allow only the numerical values of the normalized power, but it will also provide ways to diagnose other sleeping disorders. This method can be a great aid in designing brain interface system. We can also use this method to study other biomedical signals like an EKG, Electromyogram (EMG), Electrooculagram (EOG), Galvanic Skin Response, ECG, ERG etc. The Artificial neural network can also be designed using values of power spectral density of these EEG signals. Fuzzy logic is one of the fields that are still untouched for studying EEG signal. Hence, PSD estimation can be a great aid in designing Fuzzy logic.

Here, the use of Welch method is done for estimating the power spectral density. The other options for estimating power spectral density can be parametric methods like AR models, Moving average (MA) models, and autoregressive moving average (ARMA) models. Nonparametric methods like Periodogram method and Capon method can also be used in future to estimate PSD. Other filters like Butterworth, Chebyshev, and Bessel etc can also be used in filtering EEG signal. Experiments can also be done by increasing the order of filters selected in my work. EEG signals can also find application in designing human intelligence system.

REFERENCES

1. Siddiqui M. M, Srivastava G, Saeed S. H. Diagnosis of Nocturnal Frontal Lobe Epilepsy (NFLE) Sleep Disorder Using Short Time Frequency Analysis of PSD Approach Applied on EEG Signal. Biomed Pharmacol J 2016;9(1)

2. Siddiqui, Mohd Maroof, Geetika Srivastava, Syed Hasan Saeed, et al. "Detection of Rapid Eye Movement Behaviour Sleep Disorder using Time and Frequency Analysis of EEG Signal Applied on C4-A1 Channel" Communication and Power Engineering. Berlin, Boston: De Gruyter, 2016. 310-326

3. Siddiqui MM, et al. Diagnosis of insomnia sleep disorder using short time frequency analysis of PSD approach applied on EEG signal using channel ROC-LOC. Sleep Science (2016)

4. Siddiqui, Mohd Maroof, et al. "Detection of rapid eye movement behaviour disorder using short time frequency analysis of PSD approach applied on EEG signal (ROC-LOC)."Biomedical Research 26.3 (2015): 587- 593.

5. Siddiqui M. M, Srivastava G, Saeed S. H. Detection of Sleep Disorder Breathing (SDB) Using Short Time Frequency Analysis of PSD Approach Applied on EEG Signal. Biomed Pharmacol J 2016;9(1)

6. Siddiqui, Mohd Maroof, et al. "EEG Signals Play Major Role to diagnose Sleep Disorder." International Journal of Electronics and Computer Science Engineering (IJECSE) 2.2 (2013): 503-505.

7. Siddiqui, Mohd Maroof, et al. "Diagnosis of narcolepsy sleep disorder for different stages of sleep using Short Time Frequency analysis of PSD approach applied on EEG signal," 2016 International Conference on Computational Techniques in Information and Communication Technologies (ICCTICT), New Delhi, India, 2016, pp. 500-508.

8. Siddiqui, Mohd Maroof, et al. "Detection of Periodic Limb Movement with the Help of Short Time Frequency Analysis of PSD Applied on EEG Signals." Extraction 4.11 (2015).

9. Pandey, Varsha, et al. "SLEEP DISORDERS AND EEG RECORDING." International Journal of Electronics and Computer Science Engineering (IJECSE) 4.3 (2015): 206-210.

10. Akhtar, Mahnaz, Khadim Abbas, and Mohd Maroof Siddiqui. "NOCTURNAL FRONTAL LOBE EPILEPSY (NFLE): MEDICAL SLEEP DISORDER." International Conference on Emerging Trends in Technology, Science and Upcoming Research in Computer Science,DAVIM, Faridabad, 25th April, (2015):1168-1172

11. Anas, Ali, and Mohd Maroof Siddiqui. "Advent of Biometric Sensors in Field of Access Control." International Journal of Electronics and Computer Science Engineering (IJECSE) 4.3 (2015): 326-329

12. Mohd Maroof Siddiqui "Electronics Signal Help In The Treatment of Paralysis" International Journal of Electronics Signal & System (IJESS) 1.2(2012) 63-67

13. Srivastava, Sumit Kumar, Sharique Ahmed, and Mohd Maroof Siddiqui. "ANALYSIS OF BRAIN SIGNAL FOR THE DETECTION OF EPILEPTIC SEIZURE."

14. American Sleep Disorders Association (ASDA). EEG arousals: scoring rules andexamples. Sleep 15:173–184, 1992.

15. RECHTSCHAFFEN, A., KALES, A. A Manual of Standardized Terminology, Techniques and Scoring System for Sleep Stages of Human Subjects.Washington, DC: US Government Printing Office; NIH Publication No. 204,1968.

16. KRYGER, M. H., ROTH T., DEMENT, W. C. (Eds.). Principles and Practice of Sleep Medicine. Philadelphia: Saunders, 1994.

17. HALASZ, P. Hierarchy of micro-arousals and the microstructure of sleep.Clinical Neurophysiology, 28: 461-475, 1998.

18. CRISTINA MARZANO, MICHELE FERRARA, EMILIA SFORZA AND LUIGI DE GENNARO.Quantitative Electroencephalogram (EEG) in Insomnia: A New Windowon Pathophysiological Mechanisms. Current Pharmaceutical Design, 2008, 14, 3446-3455.

19. MICHAŁ SKALSKI. The Diagnosisand Treatment of Insomnia.Department of Psychiatry Medical University of Warsaw,Sleep Disorders Outpatients ClinicPoland.

20. Flynn Pharma Ltd (2012). Sleep disorders. [online] Available from www.sleepwelllivewell.co.uk/sleep-disorders.

21. Flynn Pharma Ltd (2012). Insomnia. [online] Available from www.sleepwelllivewell.co.uk/sleep-disorders/insomnia.

22. Manifestation and management of chronic insomnia in adults. Agency for Healthcare Quality and Research Web site.http://archive.ahrq.gov/clinic/epcsums/insomnsum.htm#ref2.Updated June 2005.

23. Kessler RC, Berglund P, Demler O, et al. The epidemiology of major depressive disorder: results from the National Comorbidity Survey Replication (NCS-R). JAMA. 2003;289:3095-3105.

24. Kessler RC, Berglund PA, Coulouvarat C, et al. Insomnia and the performance of US workers: results from the America Insomnia Survey. Sleep. 2011;34(9): 1161-1171.

25. American Psychiatric Association. Diagnostic and Statistical Manual of Mental Disorders.Fifth Edition.American PsychiatricAssociation Web site.http://www.psychiatry.org/practice/dsm/dsm5 .2013.

26. GAMBOA, H. Wave patterns: Delta, Theta, Alpha, Mu rhythm, Beta, Gamma[online]. 2005 Available at WWW: http://en.wikipedia.org/wiki/Electroencephalography

27. Borbély AA, Baumann F, Brandeis D, Strauch I, Lehmann D. Sleep deprivation: effect of sleep stages and EEG power density in man. ElectroencephalogrClinNeurophysiol 1981; 51: 483-93.

28. Borbély AA. A two-process model of sleep regulation. Hum Neurobiol 1982; 1: 195-204.

29. Dijk DJ, Brunner DP, Borbely AA. Time course of EEG power density during long sleep in humans. Am J Physiol 1990; 258: 650- 61.

30. Dijk DJ, Beersma DGM. Effects of SWS deprivation on subsequent EEG power density and spontaneous sleep duration. ElectroencephalogrClinNeurophysiol 1989; 72: 312-20.

31. Pigeon WR, Perlis ML. Sleep homeostasis in primary insomnia. Sleep Med Rev 2006; 10: 247-54.

32. Lichstein KI, Rosenthal TI. Insomniacs' perceptions of cognitive versus somatic determinants of sleep disturbance. J AbnormPsychol 1980; 89: 105-7.

33. Mitchell KR. Behavioral treatment of presleep tension and intrusive cognitions in patients with severe predormital insomnia. J Behav Med 1997; 2: 57-69.

34. Merica H, Gaillard JM. The EEG of the sleep onset period in insomnia: a discriminant analysis. PhysiolBehav 1992; 52: 199-204.

35. SN Arya, K Rajiv, R Singh.Practical Approach to the Diagnosis and Management of Insomnia.

36. PeraitaAdradosR.Transient and short term insomnia. In: Billiard M, ed. Sleep, Physiology and Pathology.2003; New York: Kluwer Academic/Plenum Publishers.

37. Gloor, P. 1969. Hans Berger On the Electroencephalogram of Man. The Fourteen Original Reports on the Human Electroencephalogram, N. Elsevier SciencePublishers, Amsterdam.

38. KRYGER, M. H., ROTH T., DEMENT, W. C. (Eds.). Principles and Practice of Sleep Medicine. Philadelphia: Saunders, 1994.

39. TERZANO, M. G., PARRINO, L., SMERIERI, A., CHERVIN, R., CHOKROVERTY, S., GUILLEMINAULT, C., HIRSHKOWITZ, M., MAHOWALD, M., MOLDOFSKY, H., ROSA, A., THOMAS, R., WALTERS, A. ALTAS, rules, and recording techniques for the scoring of cyclic alternating pattern (CAP) in human sleep. Sleep Medicine, 2:537–553, 2001.

40. http://physionet.org/physiobank/database/capslpdb/

41. Richard Mahlberg, Thorsten Kienast, Sven Ha¨del, Degree of pineal calcification (DOC) is associated with polysomnographic sleep measures in primary insomnia patients, Sleep Medicine, Elsevier 2008.

42. Josiah Poon, Simon Poon, Dawei Yin, Studying Herb-Herb Interaction for Insomnia through the theory of Complementarities, 2010 IEEE International Conference on Bioinformatics and Biomedicine Workshops.

43. JaapLancee, Jan van den Bout, Annemieke van Straten, Internet-delivered or mailed self-help treatment for insomnia? A randomized waiting-list controlled trial, Behaviour Research and Therapy, Elsevier 2012.

44. Vargas, Insomnia Characterisation: From Hypnogram to Graph Spectral Theory, IEEE 2015.

45. R. de León-Lomelí, J.S. Murguía, Relation Between Heart Beat Fluctuations and Cyclic Alternating Pattern During Sleep in Insomnia Patients, IEEE 2014.

46. Anna Maria Bianchi, Processing of Signals Recorded Through Smart Devices: Sleep-Quality Assessment, IEEE 2010.

47. Yue Wang, Yongchao Zhang, Yulin Huang, ADVANTAGES AND CHALLENGES OF POWER SPECTRAL DENSITY ESTIMATION METHODS FOR SCANNING RADAR ANGULAR SUPERRESOLUTION, IEEE 2015.

48. ChengzhiSu, LeiZhong, Modelling and Control for Simultaneous Laser Beam Alignment of a Dual-PSD Industrial Robot Calibration System, IEEE 2015.

49. LIU Jingwei, Deep Learning EEG Response Representation for Brain Computer Interface, 34th Chinese Control Conference, Hangzhou, China, July 28-30, 2015.

50. Mahdi Parchami, A New Algorithm for Noise PSD Matrix Estimation in Multi-Microphone Speech Enhancement Based on Recursive Smoothing, IEEE 2015.

51. Md Belal Bin Heyat, Shahabaz Ahmad Khan, Ahmad Zakariya, Faijan Akhtar and Shajan Azad, "Microcontroller Using Industrial Tank", Onyx Journal of Multi-Disciplines, Onyx InterScience, Vol. 1, Issue 1, October 2016, pp. 5-8.

52. Md Belal Bin Heyat, Shaguftah, Faijan Akhtar, Mohd Ammar Bin Hayat, Shajan Azad, "Power Spectral Density are used in the Investigation of insomnia neurological disorder", XL PRE-CONGRESS symposium, organized by INDIAN ACADEMY OF SOCIAL SCIENCES [ISSA] in King George Medical University & State Takmeel ut-tib College and Hospital Lucknow.

53. Syed Rafi Ahmed, Shahnawaz, Faijan Akhtar & Tauheed "Superiority Control of Concrete", 3rd International Seminar on Sources of Planet Energy, Environmental & Disaster Science : Challenges and Strategies (SPEEDS-2016) organized by School of Management Sciences, Lucknow, 19-20 Nov 2016.

54. Shahabaz Ahmad Khan, Ahmad Zakariya, Er Saima Beg, Md Belal Bin Heyat and Mohd Maroof Siddiqui, "Industrial Tank Temperature, Pressure and Humidity Controller Using Microcontroller", National Conference on Emerging Trends in Non Conventional Energy Resources, organized by Integral University, Lucknow, 22 October 2016.

55. Md Belal Bin Heyat, Faijan Akhtar, Shajan Azad, "A Review on use of Sunlight in Human Life", International Journal of Trend in Scientific Research and Development, Vol. 1, Issue 2, Nov- Dec 2016, pp. 22-24.

56. Md Belal Bin Heyat, Faijan Akhtar, Shafan Azad, Shadab Azad and Shaguftah, "Dual Tone Multi-Frequency Based Premises Appliance Control Switch", International Journal of Technical Research & Science, Vol. 1, Issue 7, Nov- Dec 2016, pp. 215-218.

57. Md Belal Bin Heyat, Faijan Akhtar, Mohd Sikandar Hayat Siddiqui, Shafan Azad, "An Overview of Dalk Therapy and treatment of Insomnia by Dalk Therapy", National Seminar on Research Methodology in Ilaj-Bit-Tadbeer, organized by State Takmeel-ut-Tib-College & Hospital, Lucknow 10 October 2015.

58. Y.M.Hasan, Md Belal Bin Heyat, M.M.Siddiqui, S.Azad, and F.Akhtar, "An Overview of Sleep and Stages of Sleep", International Journal of Advanced Research in Computer and Communication Engineering 2015 December; 4(12): pp. 505-507.

59. Md Belal Bin Heyat, Shaguftah, Y.M.Hasan, M.M.Siddiqui, "EEG Signals and Wireless transfer of EEG Signals", International Journal of Advanced Research in Computer and Communication Engineering Volume 2015 December; 4(12): pp. 502-504.

60. Omer Farooq, Touseef Rahman, Md Belal Bin Heyat, Mohd Maroof Siddiqui, Faijan Akhtar, "An Overview of NFLE", International Journal of Innovative Research in Electrical, Electronics, Instrumentation & Control Engg 2016 March; 4(3): pp. 209-211.

61. Touseef Rahman, Omer Farook, Md Belal Bin Heyat, Mohd Maroof Siddiqui, "An Overview of Narcolepsy", International Advanced Research Journal in Science, Engineering and Technology 2016 March;3(3):pp. 85-87.

62. Er. Shipra Srivastava, Mohd Maroof Siddiqui, Saifur Rahman, Prof (Dr.) Syed Hasan Saeed, Md Belal Bin Heyat, " Carbon Nano tubes & Its Application In Medical Field & Communication", International Journal of Advanced Research in Computer and Communication Engineering 2016 May;5(5): pp.170-173.

63. Md Belal Bin Heyat, Mohd Maroof Siddiqui, "Recording of EEG, ECG, EMG Signal", International Journal of Advanced Research in Computer Science and Software Engineering 2015 October; 5(10): pp. 813-815.

64. Md Belal Bin Heyat, Faijan Akhtar, Shaguftah, Naseem Ahmad, "An Overview of Renewable Energy", International Journal of Technical Research & Science, Volume 1, Issue 6, September 2016. pp 119-121.

65. www.physionet.org

AUTHOR'S PUBLICATIONS

- **Md Belal Bin Heyat**, Mohd Maroof Siddiqui, *"Recording of EEG, ECG, EMG Signal"*, International Journal of Advanced Research in Computer Science and Software Engineering, Volume 5, Issue 10, October-2015, 813-815.

- **Md Belal Bin Heyat**, Shaguftah, Yassir.M.Hasan, Mohd Maroof Siddiqui, *"EEG signals and wireless transfer of EEG Signals"*, International Journal of Advanced Research in Computer and Communication Engineering, Volume 4, Issue 12, December 2015, 502-504.